Table of Contents

This manual is intended to prepare technicians for the Environmental Protection Agency's (EPA) Section 608 Certification examination and contains the information required to successfully complete the exam. This book serves as a guide for reviewing material related to Section 608 of the Clean Air Act and is **not a formal refrigeration training course.** Technicians preparing for this examination should be familiar with the basic vapor-compression refrigeration cycle, as well as common service principles, practices, and procedures.

This manual has been developed with the most current information available at the time of publication. Should EPA regulations change after a technician becomes certified, it is the responsibility of the technician to comply with these changes. The EPA also reserves the right to modify the test questions and/or require new certification or recertification based on advancements in technology. The ESCO Institute will update this manual, as necessary, to reflect current EPA regulations and testing requirements.

Federal Regulations

Section 608 of the Clean Air Act requires all persons who maintain, service, repair, or dispose of appliances containing regulated refrigerants, be certified in proper refrigerant handling techniques as required by the National Recycling and Emission Reduction Program. Regulated refrigerants currently include: CFC, HCFC, HFC, and HFO refrigerants.

If you maintain, service, repair, or dispose of appliances containing a regulated refrigerant, you must be certified. **You cannot work under another person's certification.**

Before You Begin

In addition to this preparatory manual, practice questions are available to help prepare you for the EPA Section 608 examination. You can access these practice questions, free of charge, on the ESCO website at www.escogroup.org/practice.

If your examination was administered through an ESCO-approved testing location, you will be able to login to the ESCO website at www.escogroup.org to access your examination results, order replacement certification cards, update your information (i.e., address), opt out of the public certification registry, order additional training materials, etc. You may also contact our customer service team, Monday-Friday, 8:00 AM 5:00 PM, Central Time at 1-800-726-9696 if you have any questions related to your certification. **Please note**: if you participate in an examination that is administered in paper format (i.e., Scantron answer sheet) please allow 5-7 business days for your examination to be received at our grading center for processing.

Overview of the Examination

There are four (4) categories of technician certification:

Core. Required to obtain any one of the four levels of certifications.

Type I. Persons who maintain, service, repair, or dispose of small appliances must be certified as Type I technicians. A small appliance is defined as a pre-assembled unit, hermetically sealed and factory charged with 5 lbs. or less of refrigerant. Examples include equipment such as water coolers, window units, refrigerators, freezers, de-humidifiers, residential ice machines, and package terminal air conditioning. Split- systems are not included in Type I.

Type II. Persons who maintain, service, repair, or dispose of appliances, containing more than 5 lbs. of refrigerant, or if the installation of such equipment requires refrigerant charging, must be certified as Type II technicians. Type II certification does not include small appliances or motor vehicle air-conditioning (MVAC) systems.

Type III. Persons who maintain, service, repair, or dispose of low-pressure appliances, such as centrifugals and chillers, must be certified as Type III technicians. Low-pressure appliances operate with low-pressure refrigerants, which have pressures of 30 psig, or lower, at a liquid-phase temperature of 104°F. HCFC-123 used in chillers is a "low-pressure refrigerant" under EPA's Section 608 regulations to replace CFC-11.

Universal. Persons, who maintain, service or repair both low and high-pressure equipment, as well as small appliances, will be certified as Universal technicians.

Test Format

The test contains four sections: Core, and sections I, II, and III. Each section contains twenty-five (25) multiple-choice questions. A technician MUST achieve a minimum passing score of 70 percent in each group/section in which they are to be certified. For example, a technician seeking Universal certification must achieve a minimum score of 70 percent, or 18 out of 25 correct, on each section of the test. If a technician fails one or more of the sections, they may retake the failed section(s) without retaking the section(s) in which they earned a passing score. In the meantime, the technician will be certified in the Type(s) for which they received a passing score. There is one exception; a technician MUST achieve a passing score on the Core plus any one Type to receive any certification.

The Core contains 25 general knowledge questions relating to stratospheric ozone depletion, rules and regulations of the Clean Air Act, the Montreal Protocol, refrigerant recovery, recycling and reclaiming, recovery devices, substitute refrigerants and oils, recovery techniques, dehydration, recovery cylinders, safety, and shipping. Type I contains 25 sector-specific questions pertaining to small appliances. Type II contains 25 sector-specific questions pertaining to medium and high-pressure appliances and Type III contains 25 sector-specific questions pertaining to low-pressure appliances.

Federal regulations require that this exam be conducted as a closed-book exam by an authorized test administrator (Proctor). The only outside materials allowed during the test are a pressure-temperature (PT) chart and a calculator. Phones are NOT allowed to be used and MUST be turned off and put away (not on the desktop) during the examination. Use of any other electronic communication device, or attempts to copy, distribute, post publicly, share photos of exam questions, etc., may result in revocation of certification and will be reported to the U.S. EPA.

Certain personal information is requested on the exam. Technicians should be prepared to provide:

- Picture identification (Proctors will ask for this to verify your identity–**this is required**.)
- Social security number (Used for identification purposes only.)
- Home/mailing address
- Date of Birth
- Phone number
- Email address

As required, all examination participants will be included in an online registry/lookup by name, city and state, as well as certification achieved. (No personal information will be included in the public registry.) Technicians will be able to opt out of this registry by logging into the ESCO website at www.escogroup.org or by contacting customer service at 800-726-9696.

Technicians should carefully study the Core and section(s) related to the Type(s) of certification in which they are seeking to achieve a passing score. Free EPA practice exams can be found online at: www.escogroup.org/practice. Practice exams are provided to give exam candidates a sample of the types of questions they can expect to encounter on the exam and are not intended to be a sole study source.

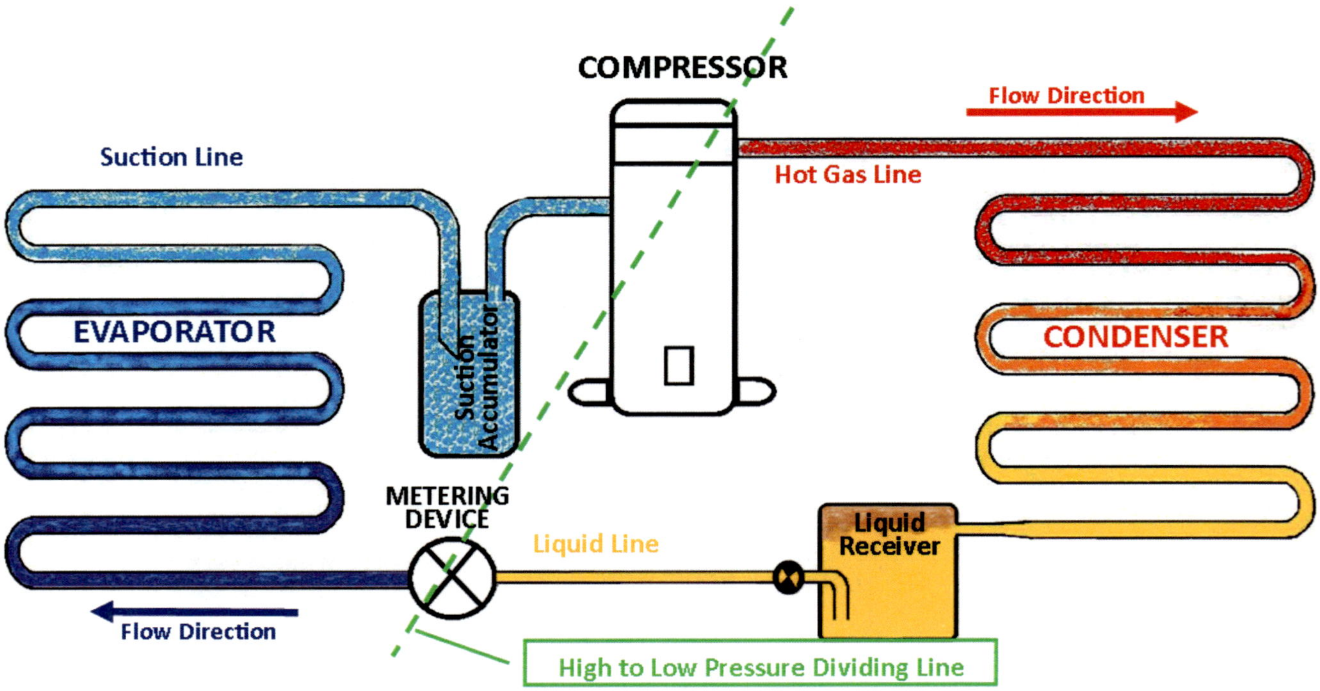

Figure 1. Vapor-Compression Refrigeration Cycle.

Vapor - Compression Refrigeration Cycle

The compressor is the heart of the vapor-compression refrigeration cycle, Figure 1. Low-pressure, low-temperature, superheated refrigerant vapor enters the compressor and is compressed, changing it to a high-pressure, high-temperature, superheated vapor. It then moves to the condenser where some of the heat is removed, de-superheating and condensing it into a liquid. Before it leaves the condenser as a high-pressure liquid, the liquid refrigerant is subcooled to a point below the liquid saturation temperature. It then flows to the metering or expansion device as a high-pressure, subcooled liquid. As the refrigerant flows through the metering device, the pressure of the liquid is reduced, causing a small percentage of the liquid to flash to a vapor (flash-gas), lowering the temperature of the remaining refrigerant to its saturation temperature. The low-pressure, low-temperature refrigerant flows into the evaporator as a low-temperature saturated refrigerant. As the refrigerant absorbs heat, it evaporates into a low-temperature vapor. This evaporation process is referred to as direct expansion. During this process, the refrigerant vapor is superheated above its saturation temperature and then enters the suction line. From the suction line, refrigerant enters the compressor as a low-pressure, low-temperature, superheated vapor to repeat the cycle. The compressor and the metering device are the dividing points between the low-pressure and high-pressure sides of the system.

Accessories shown in the basic diagram are the liquid receiver and a suction accumulator. Use of these components depends on system design and/or the type of metering device used. A system that uses a thermostatic expansion valve (TEV) is usually equipped with a receiver located in the liquid line, directly following the condenser. A system that uses a thermostatic expansion valve (TEV), capillary tube, or fixed bore metering device may be equipped with an accumulator, located in the suction line, which prevents liquid from entering the compressor.

A system may have service valves, access valves, or process stubs to gain access for service. Never front-seat (turn the valve stem clockwise as far as it will go) a service valve when the system is in operation. The valve must be back-seated (turn the valve stem counter-clockwise as far as possible) to close the service or gauge port before removing the service manifold hoses.

How to Use the Pressure-Temperature (PT) Chart

A pressure-temperature (PT) chart provides the pressure-temperature relationship of refrigerants used in the HVACR industry, Figure 2. The charts are based on the saturation properties for which a refrigerant evaporates and condenses. Liquid must be present for the relationship to match. For example, a PT chart will indicate a cylinder at 70°F containing 1 ounce or 10 lbs of liquid R-410A will have a pressure of 201 psig. Likewise, the evaporator or condensing temperature can be found by comparing the operating suction or high-side pressure for the type of refrigerant to the corresponding temperature.

F°	R-410A PSI	R-22 PSI	R-134A PSI
0°	48	24	6
20°	78	43	18
40°	118	69	35
60°	170	102	57
70°	201	121	71
80°	236	144	87
100°	318	196	124

Figure 2. Pressure-Temperature Chart.

Refrigeration Service

Refrigerant will migrate to a compressor's crankcase and mix with the oil when there is a difference between the oil pressure and refrigerant vapor pressure. The compressor crankcase heater is used to help prevent this refrigerant migration.

An oil sample should be taken when the unit has had a leak or a major component failure. Refrigerant oil samples are also taken when moisture, acid, oil sludge formation, oil waxing, or residual acidic oil from a burn-out are suspected problems.

Finding and repairing leaks in the system will conserve refrigerant when servicing an appliance. Flushing field tubing with liquid refrigerant to clean the tubing after a burn-out is not recommended, and it is unlawful.

After making a major repair, the appliance should be dehydrated by evacuating the system to a minimum of 500 microns. This minimum level is required for HFC refrigerants with POE oil or any other refrigerant system. Under no circumstance should a hermetic compressor be operated when there is a vacuum in the system; electrical arcing could burn the terminals inside the compressor.

Gauge Manifold Set

One of the most important tools for an HVACR technician is the gauge manifold set, Figure 3. The left side, low-pressure, compound gauge (blue) and the right side, high-pressure gauge (red), are attached to the manifold to measure system pressures. Hoses are used to connect the manifold to the refrigeration system's access ports, or service valves, to gain access to system pressures. The hoses are also color-coded to match the gauges.

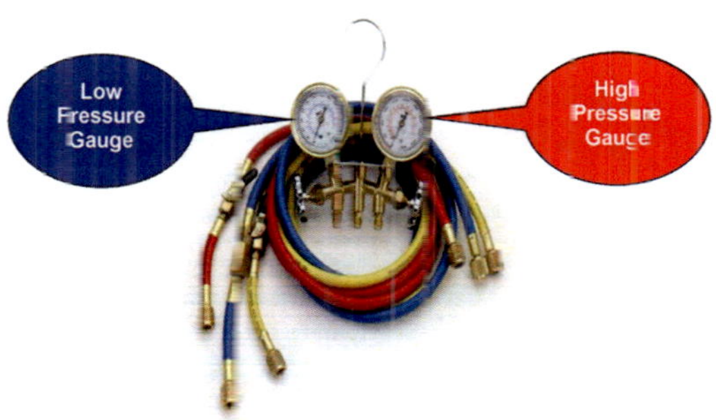

Figure 3. Gauge Manifold Set.

The compound pressure gauge measures system pressure for the low side in pounds per square inch gauge (psig) and vacuum in inches of mercury (Hg). The high-pressure gauge measures high-side (discharge) pressure. Most high-pressure gauges range from 0 to 800 psig. The manifold is also equipped with a center port, (usually a yellow hose), that can be used for recovery, evacuation, and charging connections.

An electronic manifold may have combined temperature probes to measure refrigerant line temperatures for calculating system superheat and subcooling, Figure 4.

Figure 4. Digital Manifold & Temperature Kit.

The EPA recommends that hoses be equipped with low-loss fittings or valves that manually close, or which close automatically, to minimize refrigerant loss when hoses are disconnected. The hoses used for service and with recovery equipment must be equipped with low-loss fittings.

A minor release of refrigerant, when connecting or disconnecting hoses for service or recovery, is considered a de-minimis release and is not considered unlawful.

Caution: The gauge manifold and hoses must be pressure-rated to handle the refrigerant being used.

Tips for Taking Multiple-Choice Exams

The EPA Section 608 exam is a multiple-choice examination. Please review the strategies listed below to help you perform your best on the examination.

Be sure to review your preparatory manual. Whether you have taken a class or not, it is still a good idea to review your preparatory manual. There may be information in this book that was not reviewed in class.

Try taking practice exams. While practice questions will not provide you with the exact questions you will see on your exam, they will help you prepare for taking a multiple-choice exam and give you an idea of the content you may see on the exam.

Read the entire question before selecting an answer. Reading a question thoroughly before reading and selecting the answer choices will ensure you better understand the question.

Read all answer choices before choosing an answer. Be sure to read all answer choices before selecting an answer. Sometimes what appears to be correct may not be once you read all the answer choices.

Answer questions you know first. Read through and answer the questions you know so you can go back and concentrate on the questions you are unsure of.

If you are unsure of an answer, use the process of elimination. Read through and eliminate the answer choices you are 100 percent sure are not correct. This will help you limit the answer choices.

Stick with your first answer choice. Your first answer choice is *usually* correct. (Not always, but usually.) Try not to second guess yourself and change all your answers.

Core: Required for any one of the four levels of certification.

Stratospheric Ozone Depletion

Stratospheric ozone helps form the earth's protective shield. The ozone layer protects the earth from ultraviolet radiation from the sun. Ozone depletion potential (or ODP) measures the ability of a substance to destroy ozone in the stratosphere. Ozone depletion in the stratosphere is a global problem which affects everyone on the planet. One of the most serious results from the thinning and damaging of the stratospheric ozone layer is an increase in the rates of skin cancer in humans. Another human health effect that has increased from damage to the stratospheric ozone layer is the increase in cases of cataracts. Stratospheric ozone depletion can also have an effect on the environment, causing decreased crop yields and damage to marine organisms.

There has been some controversy over the subject of ozone depletion. Some believed that the chlorine found in the stratosphere comes from natural sources such as volcanic eruptions. However, air samples taken over erupting volcanoes show that volcanoes add only small quantities of chlorine to the atmosphere compared to the amount of chlorine added to the atmosphere from chlorine-containing refrigerants. In addition, the rise in the amount of chlorine measured in the stratosphere over the past four decades matches the rise in the amount of Fluorine, which has different natural sources than chlorine over the same period. Also, the rise in the amount of chlorine measured by NASA and other agencies in the stratosphere over the past twenty years matches the rise in CFC and HCFC emissions over the same period. The evidence is clear, chlorine containing refrigerants have changed the natural balance thus depleting the ozone layer.

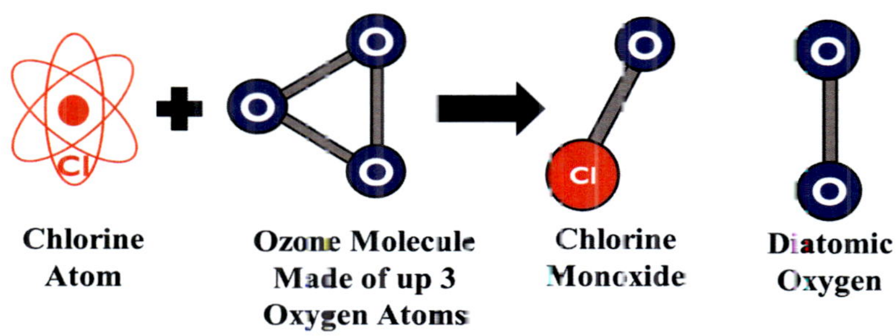

Figure 5. Chlorine / Ozone.

Ozone in the stratosphere consists of molecules containing 3 oxygen atoms, O_3. Chlorine is the element of a CFC or HCFC molecule that causes ozone depletion. When a chlorine atom encounters an ozone molecule, it takes one of the ozone's 3 oxygen atoms, forming a compound called chlorine monoxide (ClO), and leaves an oxygen O_2 molecule behind, Figure 5. The chlorine monoxide will collide with another ozone molecule, releasing its Oxygen atom, forming two O_2 molecules, leaving the chlorine free to attack another ozone molecule. A single chlorine atom can last in the Stratosphere for 120 years destroying up to 100,000 ozone molecules.

Unlike other chlorine compounds and naturally-occurring chlorine, the chlorine in CFCs and HCFCs will neither dissolve in water nor break down into compounds that dissolve in water so they do not rain out of the atmosphere.

HFCs are made up of Hydrogen, Fluorine, and Carbon. They do not contain chlorine that effects the ozone layer, but most HFCs have a high Global Warming Potential (GWP).

Global Warming Potential

Global Warming Potential (GWP) was established to provide comparisons of the global warming impacts of different gases over a span of time. Carbon Dioxide (CO_2) is used as the baseline measurement for global warming potential and has a value of 1. Most HFC refrigerants, such as R-410A, have global warming potentials thousands of times greater compared to carbon dioxide. Hydrocarbon (HC) and Hydrofluoroolefin (HFO) refrigerants have the lowest global warming potential compared to CFCs, HCFCs, and HFCs.

The global warming potential of isobutane (R-600a), propane (R-290), and R-441A are significantly lower when compared to the GWPs of HFCs such as R-134a, R-407C, R-404A, and R-410A.

With non-ozone depleting and very low global warming potential characteristics, hydrofluoroolefin (HFO) refrigerants are a good fit for HVACR equipment. Most HFCs have very high global warming potentials; a characteristic that makes them damaging to the environment.

Refrigerant Characteristics & Identification

Refrigerants may be identified by type (CFC, HCFC, HFC, HFO, HC) and American Society of Heating Refrigerating and Air-Conditioning Engineers (ASHRAE) number or with "R" plus the ASHRAE number such as HFC-134a or R-134a.

ASHRAE has grouped refrigerants by Class A (safest) or B, depending on their toxicity level to humans. Flammability is indicated by a 1 (no flammability), 2 (low flammability), or 3 (high flammability), Figure 6.

Safety Group

Flammability (Low → Higher)	Toxicity: Low	Toxicity: Higher
High Flammability	A3	B3
Low Flammability	A2	B2
Slight Flammability	A2L	B2L
No Flammability	A1	B1

Figure 6. ASHRAE Safety Group.

Chlorofluorocarbon (CFC) refrigerants have the highest ozone depletion potential (ODP) and are the most harmful to stratospheric ozone. CFC refrigerants contain chlorine, fluorine, and carbon. Chlorine is the component that poses a threat to the stratospheric ozone. Examples of CFC refrigerants include R-11 and R-12.

Hydrochlorofluorocarbon (HCFC) refrigerants contain hydrogen, chlorine, fluorine, and carbon. The hydrogen makes them a little less harmful to the stratospheric ozone when compared to CFCs. R-22 and R-123 are examples of HCFC refrigerants that have an ozone depletion potential of more than zero, ranging between zero and one.

Hydrofluorocarbon (HFC) refrigerants contain hydrogen, fluorine, and carbon atoms connected by single bonds between the atoms. HFCs have the lowest/no ozone depletion potential but do have an effect on global warming. HFC refrigerants do not contain harmful chlorine. Examples of HFCs are R-134a, R-410A, R-404A, R-407C, and R-422B.

Hydrofluoroolefin (HFO) refrigerants contain hydrogen, fluorine, and carbon atoms connected with at least one double bond between the carbon atoms. HFOs have no ozone depletion potential but do have a very small effect on global warming. HFOs are less flammable than hydrocarbon refrigerants but most are still mildly flammable so are classified as A2L. HFO refrigerants contain fluorine making them less flammable than hydrocarbon refrigerants. HFO refrigerants are miscible in POE lubricants. R-1234yf is an HFO refrigerant used in chillers.

Hydrocarbon (HC) refrigerants are an elementary compound of hydrogen and carbon. Although flammable, these refrigerants pose the least amount of danger to the environment. HCs have no ozone depletion potential and a very small effect on global warming, with GWP values less than 10. Isobutane (R-600a) and Propane (R-290) are examples of HCs. Some HC refrigerants are considered natural refrigerants.

Note: Propane cylinders for grilling should never be used as a refrigerant, as they contain impurities that can damage refrigeration equipment.

Azeotropic refrigerant mixtures contain two or more refrigerants. At a certain pressure, an azeotropic mixture evaporates and condenses at a constant temperature. It acts like a single component refrigerant over its entire range. Thus, azeotropic mixtures act as a pure compound.

Zeotropic or **non-azeotropic** refrigerant mixtures are mixtures of components that have different boiling points. This mixture is a blend of refrigerant. The blend may be binary (two-part) or ternary (three-part). Blended refrigerants are normally associated with having what is called "temperature glide". This is due to the blended parts having different pressures for the same saturation temperature. Temperature glide can range a few tenths of a degree to 12 degrees or more.

Bubble and Dew Points

Zeotropic and some near-azeotropic blends use bubble and dew points to indicate condensing and evaporation temperatures on a pressure-temperature chart. Bubble point (liquid) values are used when charging by condenser subcooling. Dew point (vapor) values are used for charging by suction or evaporator superheat. The relationship of pressures and temperatures are based on the refrigerant's temperature glide.

The R-400 series of refrigerants are near azeotropic blends that can leak from a system at uneven rates due to different vapor pressures which can affect the percentage of each refrigerant remaining in the system. The proper charging method for R-400 series blended refrigerants is to weigh the refrigerant into the high side of the system as a liquid. When adding refrigerant to an undercharged system, liquid refrigerant is throttled into the low side with the system operating.

There is concern for using blended refrigerants in systems that develop a leak. There may be fractionization of the refrigerant (refrigerant leaking at uneven rates due to different vapor pressures). R-407C is one of the refrigerants that has a high-temperature glide and may easily experience fractionation. Even though the pressure-temperature relationship and operating characteristics may be almost the same, the refrigerants cannot be interchanged. Although the saturation pressure-temperature behavior of two refrigerants might be similar, potential of the refrigerants to oxidize copper tubing must also be considered. Due to flammability of some HCs and HFOs, equipment must be designed to handle the refrigerant. Therefore, EPA regulates the maximum amount of refrigerant that can be used in the new systems by their type and use.

When retrofitting or converting a system to use a different refrigerant, only EPA-approved substitutes can be used. Remember, EPA does not approve any substitute refrigerants as direct drop-in replacements to service an R-22 or any other system.

Low, Medium, High, and Very High Pressure

To better define leak rates and service recovery levels the EPA lists refrigerants in four liquid-phase pressure ranges at 104°F condensing temperature.

Low-pressure refrigerants have a pressure of 30 psig or lower at a liquid-phase temperature of 104°F.
Examples: CFC-11, HCFC-123, HFC-245fa, HFO-1233zd

Medium-pressure refrigerants have a pressure between 30 psig & 155 psig at a liquid-phase temperature of 104°F.
Examples: CFC-12, HCFC-124, HCFC-409A, HFC-134a, HC-600a, HFO-1234yf, HFO-1234ze

High-pressure refrigerants have a pressure between 155 psig & 340 psig at a liquid-phase temperature of 104°F.
Examples: HCFC-22, HFC-404A, HFC-407C, HFC-422B, HFC-422D, HFC-410A, HC-441A, R-717

Very high-pressure refrigerants have a pressure over 340 psig at a liquid-phase temperature of 104°F.
Examples: CO_2 (R-744)

Refrigerant Oils

Most HCFC binary (two-part) and ternary (three-part) blends use a synthetic alkylbenzene lubricant. When retrofitting HCFC equipment to HFC refrigerants or when working with newly manufactured systems, synthetic polyolester oil is most commonly used. Ester-based oils cannot be mixed with any other type of oil. Before retrofitting an HCFC system to HFC refrigerant, the other types of oils must be removed from the system before adding the ester-based oil. Care must be taken when using ester-based oils; they are very hygroscopic and will absorb moisture out of the air and through some containers.

Montreal Protocol and Clean Air Act

The United States is part of the Montreal Protocol which is an international treaty that addresses ozone-depleting substances and their alternatives. To stop damage to the stratospheric ozone layer, the U.S. is phasing out, or has already phased out, the use of chlorofluorocarbons (CFCs), hydrochlorofluorocarbons (HCFCs), and other refrigerants that are harmful to the environment.

Section 608 of the Clean Air Act sets forth the rules and regulations for refrigerant handling in the HVACR industry. Section 608 certification allows a technician to purchase or handle regulated refrigerants. Distributors that sell HCFC, HFC, or another regulated refrigerant must verify the purchaser is, or employs, a Section 608 certified technician.

Anyone performing any activity that violates the refrigerant circuit, including removal of refrigerant or replacing a component containing a regulated refrigerant must be certified at the proper level and have the type of recovery equipment required. It is illegal to knowingly release CFC, HCFC, or HFC refrigerants during the service, maintenance, repair, or disposal of appliances. A technician must have their Section 608 certification card in their possession when preforming an activity regulated by Section 608. If a technician's Section 608 certification card is lost, a replacement can be requested from their certifying organization, not the EPA.

A person is not required to be a Section 608 certified technician when servicing an external electric circuit or sections of the system that do not involve the refrigerant components of an HVACR appliance.

The Clean Air Act (EPA's regulations) allows for releasing an exempt refrigerant (e.g., ammonia and CO_2) or a small amount of refrigerant, mixed with nitrogen (a leak trace gas) for leak detection. As of 2017, a technician can be fined $44,539 per day, per violation, for violating the Clean Air Act, including knowingly releasing non-exempt refrigerant from appliances. Service technicians who violate Clean Air Act provisions may lose their certification in addition to paying fines and being required to appear in court. It is the technician's responsibility to comply with any future changes in the law or EPA regulations after a technician becomes certified. State and local governments may establish laws that contain stricter requirements than the Clean Air Act/EPA regulations.

The EPA's Significant New Alternatives Policy (SNAP) Program identifies refrigerants with lower overall risks to human health and the environment.

The phase-out for the production and imports of newly manufactured R-22 and HCFC-142b in the U.S. will be in the year 2020. After the phase-out of all HCFC refrigerants in 2030, supplies of those refrigerants for equipment servicing will come from recovered and reclaimed gas. It is not a violation of the refrigerant management regulations under the Clean Air Act to service a system that uses CFCs or HCFCs after the phase-out of those refrigerants.

Reclaimed refrigerant must meet AHRI Standard 700 before it can be resold under EPA's regulations. Technicians can only charge used CFC, HCFC, or HFC refrigerants back into the same appliance or into another appliance owned by the same owner. Used refrigerant may no longer meet the AHRI Standard for virgin refrigerant so it cannot change ownership without being reclaimed.

All CFC, HCFC, HFC, and HFO refrigerant recovery equipment must meet EPA standards. Recovery equipment must be tested according to AHRI Standard 740. Recovery machines may be able to recover one or multiple types of refrigerants.

Desiccant dehumidifiers are NOT covered by EPA Section 608 regulations as they do not contain a refrigerant.

Self-contained recovery devices remove refrigerant from an appliance without the assistance of components in the appliance. Recovery equipment that relies on the compressor in the appliance and/or the pressure of the refrigerant in the appliance is considered to be system-dependent.

To make it easier to recover refrigerant, EPA regulations require appliances that do not have service valves to have a service aperture or process stub.

When disposing of appliances with 5-50 pounds of refrigerant, EPA does not require technicians (or the company employing the technician) to record the model and serial number of the appliance. However, records that must be kept include;

- type of refrigerant recovered
- quantity of refrigerant, by type
- quantity recovered from disposed appliances in each calendar month
- quantity of refrigerant, by type, sent for reclamation or destruction.

EPA's refrigerant management regulations exempt certain refrigerants from the venting prohibition. This can occur when and if the EPA determines that a certain refrigerant does not pose a threat to the environment if released. Some natural refrigerants, such as carbon dioxide, can be purchased by someone who is not Section 608 certified.

It is lawful to service existing R-22 systems and use existing supplies of R-22. It is not legal to top off an R-22 charge with another refrigerant such as R-410A. Topping off one type of refrigerant with another can void equipment warranty, ruin the refrigerant, damage the equipment, and cause severe personal injury, or death.

When discarding any disposable cylinders of CFCs, HCFCs, HFCs, or HFOs, recover all remaining refrigerant, render the cylinder useless, and recycle the metal.

Before opening or disposing of any appliance containing a chlorofluorocarbon (CFC), hydrochlorofluorocarbon (HCFC), or hydrofluorocarbon (HFC) refrigerant, recover the refrigerant to the approved EPA recovery or evacuation levels. Under the Section 608 "Safe Disposal Requirements", it is the final person in the disposal chain that is responsible for ensuring that any CFC, HCFC, or HFC refrigerant has been removed from household refrigerators or other appliances before they are disposed of.

The knowingly release of a CFC, HCFC, HFC or HFO refrigerant during the service, maintenance, repair, vandalism, theft, or disposal of appliances is illegal. Refrigerants emitted when cutting a line without properly evacuating the appliance violates the venting prohibition. Venting refrigerant from a recovery machine or recovery cylinder after use is also illegal. Records of recovered refrigerant must be kept to ensure venting does not take place after recovery from an appliance.

Small quantities of isobutane, used in household freezers, can be vented. Releasing mixtures of nitrogen and refrigerant that result from adding nitrogen to a fully charged appliance to leak check the appliance is considered a violation of the prohibition on venting. Only a few ounces of refrigerant mixed with nitrogen is considered a leak trace gas and may be released without recovery.

Recovery

The processes of recovery, recycling, and reclaiming sound similar, but they are quite different.

Recover means to remove refrigerant, in any condition from a system and store it in an approved recovery cylinder or container. Recovered refrigerant may be charged into the same system, another system with the same ownership, or shipped to an approved reclaimer.

Recycle means to extract refrigerant from an appliance (except motor vehicle air conditioners or MVACs) and clean the refrigerant for reuse in equipment of the same owner without meeting all of the requirements for reclamation. In general, recycled refrigerant is refrigerant that is cleaned using oil separation and single or multiple passes through devices, such as replaceable core filter-driers, which reduce moisture, acidity, and particulate matter. Recycled refrigerant may be charged into the same system or another system with the same ownership. Recycling equipment must meet AHRI 740 Performance Rating Standard.

Reclaim means to process refrigerant to a level equal to new (virgin) product specifications as determined by chemical analysis. Reclaimed refrigerants must meet AHRI 700 Standards, before it can be sold. This is normally preformed at a chemical processing company.

Recovery Devices

Refrigerant recovery and/or recycling equipment manufactured after November 15, 1993 must be certified and labeled by an EPA-approved equipment testing organization to meet EPA standards. An EPA-approved certification label is required on all new recovery equipment.

There are two basic types of recovery devices: "System-dependent" (passive) which captures refrigerant with the assistance of components in the appliance from which refrigerant is being recovered, Figure 7. "Self-contained" (active) which has its own means to draw the refrigerant out of the appliance, Figure 8.

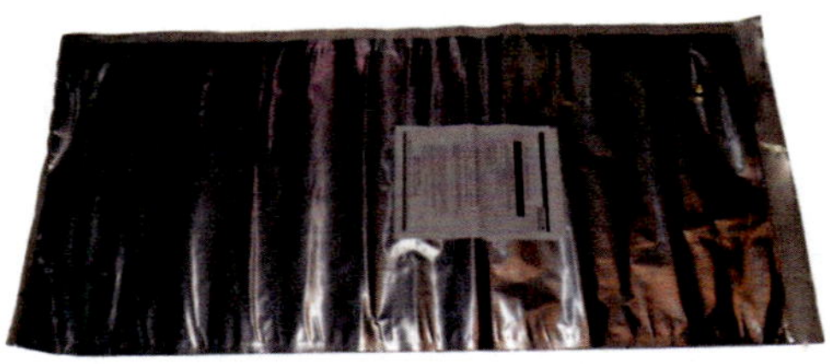

Figure 7. Passive Recovery Bag 12 oz or less.

Figure 8. Active Recovery Unit. Courtesy of Appion Inc.

The recovery time will be increased if the appliance is located in low ambient temperatures or long hoses are used between the unit and the recovery machine. Long and small diameter hoses between the unit and recovery machine should be avoided as they will cause excessive pressure drop and increase recovery time. Upon completion of refrigerant liquid transfer between recovery unit and the refrigeration system, guard against trapping liquid refrigerant in the service hose between closed service valves.

The technician must have a separate refrigerant recovery cylinder for each type of refrigerant recovered and should have a separate cylinder for refrigerants known to be mixed. If servicing systems that use HFC-134a, HFC-410A and HCFC-22, then three cylinders are required even though two are HFCs. It is important not to mix refrigerants in the same container when recovering as refrigerants that are mixed be may be impossible to reclaim and there may be an added cost to have them destroyed.

Technicians can only charge used CFC, HCFC, or HFC refrigerants back into the same appliance or into another appliance with the same owner. Used refrigerant may no longer meet the AHRI standard for virgin refrigerant so it cannot change ownership without being reclaimed.

Leak Detection

Leak detection methods and devices used must be approved by the manufacturer. The wrong type of electronic leak detector used with a flammable refrigerant can cause an explosion and bodily injuries.

To determine the general area of a small leak, the use of an electronic or ultrasonic leak detector is considered to be the most effective. To cause the least amount of damage to the environment, newly installed systems should be pressurized with dry-nitrogen for leak detection purposes. If an electronic leak detector is to be used, a trace of the system refrigerant can be added to the system along with the dry-nitrogen. The trace gas should be the same as the refrigerant the appliance is going to be charged with. A leak trace gas is not considered a refrigerant under the EPA's refrigerant management regulations. Once the general area of the leak is located, the use of soap bubbles will aid in pinpointing the leak. Finding and repairing leaks in a system will conserve refrigerant for future use.

Upon completion of a leak repair, and before recharging the system, install a new filter drier, and complete a standing-pressure leak check at the maximum system pressure. For safety, any system designed to contain a flammable hydrocarbon (HC) or HFO refrigerant has to have a leak test before evacuating to 500 microns or lower.

Unlike natural gas, flammable hydrocarbon or HFO refrigerants do not contain odorants to indicate their presence. Odorants do not need to be added to systems containing flammable refrigerants.

Dehydration

The reason for dehydrating a refrigeration system is to remove water and water vapor, and it is important to follow proper dehydration procedures. If moisture is allowed to remain in an operating refrigeration system, hydrochloric and hydrofluoric acids may form. To dehydrate a system, it should be evacuated. It is not possible to over-evacuate a system. Most manufacturers require system evacuation to 500 microns, or lower, which can be measured with a micron gauge, Figure 9.

Figure 9. Micron Gauge. Courtesy of Appion Inc.

Never evacuate a system to the ambient air without first following proper recovery procedures and attaining the mandated recovery level required by EPA regulations.

The factors that affect the speed and efficiency of evacuation are; size of equipment being evacuated, ambient temperature, amount of moisture in the system, the size of the vacuum pump and vacuum lines, Figures 10 & 11. In addition, vacuum lines (hoses) should be equal to or larger than the pump intake connection. The brand of a micron gauge does not usually affect the speed of evacuation, but the method used to connect it to the system may. The piping connection to the vacuum pump should be as short in length as possible and as large in diameter as possible. For accurate readings during evacuation, the system vacuum gauge should be connected as far away from the vacuum pump as possible. Measuring a final system vacuum should be done with the system isolated and the vacuum pump turned off. A system that will not hold a vacuum after it has been evacuated probably has a leak. During evacuation, you may wish to heat the refrigeration system to decrease dehydration time. Tapping a compressor with a rubber mallet will aid in releasing trapped refrigerant from the oil.

Figure 10. Vacuum Pump. Courtesy of Appion Inc.

Figure 11. Vacuum Manifold. Courtesy of Appion Inc.

Dehydration is complete when the vacuum gauge/indicator shows that you have reached and held the required vacuum of 500 microns or less. If the pressure increases then stops for a few minutes then rises more, there is probably moisture left in the system. If the pressure steadily rises, there is probably a leak.

Safety/General

The EPA is concerned not only with the prevention of refrigerant venting, but with the technician's overall safety. When handling and filling refrigerant cylinders or operating recovery or recycling equipment, you should wear safety glasses, protective gloves, and follow all equipment manufacturer's safety precautions. Make sure your recovery machine is grounded when in use, especially when recovering a flammable refrigerant.

When pressurizing a system with nitrogen, you should always charge through a pressure regulator and a relief valve in the downstream line from the pressure regulator. When using more than one pressure relief valve, they MUST NOT be installed in series TO EACH OTHER, only parallel to each other. If corrosion build-up is found within the body of a relief valve, the valve MUST be replaced.

When leak checking a system, NEVER pressurize the system with oxygen or compressed air. When mixed with some refrigerants or compressor oil, oxygen or compressed air can cause an explosion. To determine the safe pressure for leak testing, check the equipment data plate for the maximum low-side test pressure value.

NEVER expose refrigerants to open flames or glowing hot metal surfaces. Although reclaimers may accept visibly burned recovery tanks, at high temperatures, refrigerants decompose and form acids. Hydrochloric acid is formed if the refrigerant contains chlorine. Hydrofluoric acid is formed if the refrigerant contains fluorine. Heating a refrigerant cylinder can result in an explosion and cause serious injuries. If oxygen is also present, it is possible to form carbon monoxide, carbon dioxide, and phosgene gas. If a hydrocarbon (HC) refrigerant is released into a space, and an ignition source is provided an explosion can occur if the refrigerant concentration is above the Lower Flammability Limit (LFL) and below the Upper Flammability Limit (UFL) with an ignition source.

When working with any solvents, chemicals, or refrigerants, Safety Data Sheets (SDS) should be reviewed by the technician to reference compatibility with other materials. Silicone elastomers, for example, which are used in seals and gaskets, are not compatible with HFO refrigerants.

Refrigerant vapors, or mist, in high concentrations, should not be inhaled; they may cause heart irregularities or unconsciousness. In most refrigerant accidents where death occurs, oxygen deprivation is the major cause. According to the ASHRAE refrigerant safety classification standard, A1 designations would be the safest. R-410A is considered to be non-toxic but it, as any other refrigerant in high concentrations, can cause asphyxia.

Prior to recovery of any flammable refrigerant ensure that the refrigeration system and recovery unit is grounded. Never apply an open flame or live steam to a flammable refrigerant cylinder, do not cut or weld any refrigerant line while refrigerant is in the unit, and do not use oxygen to purge lines or to pressurize the machine. A sufficient concentration of the refrigerant in the space and an ignition source are most likely to lead to an explosion with refrigerants. HFO-1234yf is classified A2L not as an A3 (higher flammability) refrigerant like Isobutane (R-600a), Propane (R-290), and R-441A (Hydrocarbon blend).

A red color marking is required on all process tubes and other pipes through which a flammable refrigerant flows/passes and where a service connection is probable. The red marking must extend a minimum of one inch in both directions from such locations.

In the event of a large release of CFC, HCFC, HFC, HFO, HC, or any refrigerant in a contained area, a self-contained breathing apparatus (SCBA) or leaving the area is required.

Note: An alcohol spray should be used to remove ice from sight glasses or viewing glasses.

Safety/Cylinder

A refrigerant recovery cylinder should be free of rust, damage, be labeled, secure, and filled to no more than 80% of its capacity by weight. Approved refrigerant recovery cylinders can be identified by having yellow colored tops and gray bodies, Figure 12. One should not heat a refrigerant storage or recovery tank with an open flame because refrigerant in the tank may decompose forming a toxic material, it can result in venting refrigerant to the atmosphere from the pressure safety valve or the tank may explode, causing serious injury.

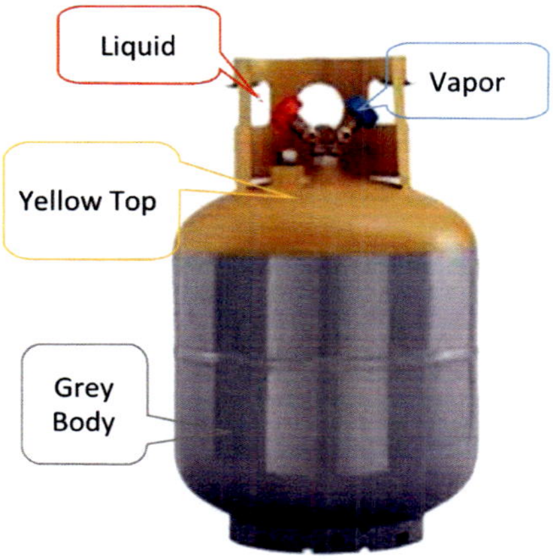

Figure 12. Recovery Tank.

When using vapor or liquid recovery, the fill level of the recovery cylinder can be controlled by a mechanical float device, electronic shut-off device, or measuring the gross cylinder weight. The pitch of the recovery device does not control the fill level of the recovery cylinder. Care must be used to prevent overfilling refrigerant storage cylinders. The internal pressure of the cylinders may rise in heated areas and cause an explosion. A reusable container for refrigerant that is under high pressure (above 15 psig), at normal ambient temperature, must be hydrostatically tested every 5 years.

A disposable refrigerant cylinder may **never** be used to recover refrigerant. When empty, the vapor should be recovered, reducing the pressure in a disposable cylinder to 0 psig before scrapping.

Shipping

When transporting cylinders containing used refrigerant a Department of Transportation (DOT) classification tag and label must be attached to the cylinders. Portable, refillable tanks or containers used to ship CFC, HCFC, or other refrigerants obtained with recovery equipment must meet DOT standards. DOT regulations require that the number of cylinders of each gas be recorded on the shipping paper for hazard class 2.2 Nonflammable Compressed Gases. Refrigerant cylinders must be positioned upright when they are shipped.

Before shipping any used refrigerant in a cylinder, a refrigerant label to identify the type of refrigerant recovered is placed on a recovery cylinder to avoid accidental mixing of recovered refrigerants, allow the recycler to identify the contents, and allow the technician's company to determine the amount of refrigerant recovered for recordkeeping purposes.

The following table lists refrigerants with characteristics that may be covered in each of the sections; Core, Types I, II & III. This table is provided for reference/comparison and does not need to be fully memorized for the exam.

Ref. No.	Type	Name/Composition	Ozone Depletion Potential (ODP)	Global Warming Potential (GWP)	Common Use	EPA Pressure Range	ASHRAE Class
R-11	CFC	Trichlorofluoromethane	1 (Base)	4000	Air-conditioning Chillers (Phased Out)	Low	A1
R-12	CFC	Dichlorodifluoromethane	1	10900	Automotive, Domestic, Commercial (Phased Out)	Medium	A1
R-22	HCFC	Monochlorodifluoromethane	0.055	1810	Air-conditioning	High	A1
R-123	HCFC	Dichlorotrifluoroethane	0.012	77	Commercial Chillers	Low	B1
R-502	CFC + HCFC	Monohloropentafluoroethane/ Monohlorodifluoromethane	.283	4657	Low-temperature Commercial	High	A1
R-134a	HFC	Tetrafluoroethane	0	1430	Automotive, Domestic, Commercial	Medium	A1
R-404A	HFC	Pentafluoroethane, Trifluoroethane, Tetrafluoroethane	0	3920	Medium and low-temperature commercial and industrial	High	A1
R-407C	HFC	Difluoromethane, Pentafluoroethane, Tetrafluoroethane	0	1770	Air-conditioning R-22 retrofit	High	A1
R-410A	HFC	Difluoromethane, Pentafluoroethane	0	2090	Air-conditioning	High	A1
R-422B	HFC	Pentafluoroethane, Trifluoroethane, Isobutane	0	2530	Retrofit for medium & low temp., R-22,R-407C, R-502 & other HCFC blends	High	A1
R-441A	HC	Ethane, Propane, Butane, Isobutane	0	<5	Domestic Refrigerators & Freezers, Vending machines, & Self-contained Commercial Refrigerators & Freezers	High	A3
R-170	HC	Ethane	0	6	Manufacturing	Very High	A3
R-290	HC	Propane	0	3	Commercial & Industrial Process Refrigeration	High	A3
R-600a	HC	Isobutane	0	3	Domestic Small Appliances, Commercial & Industrial Process Refrigeration	Medium	A3
R-744		Carbon Dioxide	0	1 (Base)	Commercial & Industrial Process Refrigeration	Very High	A1
R-717		Ammonia	0	0	Air-conditioning, Commercial & Industrial Process Refrigeration	High	B2L
R-1234yf	HFO	Hydrogen, Fluorine, Carbon	0	4	Automotive	Medium	A2L
R-1234ze	HFO	Hydrogen, Fluorine, Carbon	0	6	New Chillers, heat pumps, and vending	Medium	A2L
R-1233zd	HFO	Hydrogen, Fluorine, Carbon	0	4.7-7	New Chillers & Foam blowing	Low	A1

Type I

Persons handling refrigerant during the maintenance, service, or repair of small appliances must be certified as either a Type I Technician or as a Universal Technician, with the exception of motor vehicle air conditioning (MVAC).

The EPA definition of a small appliance includes; products manufactured, fully charged, and hermetically sealed in a factory containing five pounds or less of refrigerant. Split-systems may not be serviced by Type I technicians. The sale of CFC, HCFC, HFC, and HFO refrigerants to service or install refrigeration and air-conditioning equipment is restricted to technicians who are EPA certified in refrigerant recovery.

MVAC-like systems do not meet the criteria for Type I appliance certification no matter the type or quantity of refrigerant. HCFCs and HFCs such as R-404A cannot be used for new Type I appliances.

When retrofitting or converting a system to use a different refrigerant, only EPA-approved substitutes can be used. Remember, there is no such thing as a “drop-in” substitute; EPA does not approve any substitute refrigerants as a direct drop-in replacement to service an R-22 or other system.

Recovery Equipment

Recovery equipment used during the maintenance, service, or repair of small appliances that use CFCs, HCFCs, and HFCs must be certified by an EPA-approved laboratory.

Persons who open or perform maintenance, service, or repair of Type I appliances containing a CFC, HCFC, or HFC, can use either self-contained (active) or passive recovery equipment. A domestic refrigerator is an example of an appliance on which it would be permissible to use a passive recovery device. System-dependent or passive recovery equipment use for an appliance containing HCFCs or HFCs is limited to a maximum charge of 15 pounds.

Maintenance practices for recovery equipment should include checking for refrigerant leaks and oil levels on a regular basis. A refrigerant scale or tank internal float device are acceptable methods for monitoring that the fill level in a recovery tank does not exceed 80%.

Self-contained recovery devices, for use with small appliances containing CFCs, HCFCs, HFCs, and HFOs, must be able to recover 90% of the refrigerant when the compressor is operating or achieve a 4-inch vacuum under the conditions of AHRI 740. With an inoperative compressor, the equipment used to recover CFC, HCFC, and HFC refrigerants from small appliances, for the purpose of disposal, must be able to recover 80% of the refrigerant or achieve 4 inches of vacuum.

Refrigerant recovery devices used to recover CFCs, HCFCs, HFCs and HFCs can be either manually closed or automatically closed when disconnected to prevent loss of refrigerant from hoses. When attaching a gauge set to check system pressures, hoses with manual or self-sealing valves on the ends should be used to minimize the release of refrigerant.

When filling a charging cylinder with a regulated refrigerant, the refrigerant that is vented off the top must be recovered.

A vacuum pump made for system dehydration may be used to evacuate a recovery cylinder to be used as a recovery device, however it cannot be used in combination with a pressurized recovery cylinder or vessel for recovery.

Recovery Requirements

Hydrocarbon (HC) R-600a is an approved refrigerant for new household refrigerators, freezers, and combination refrigeration/freezers. Hydrocarbon refrigerants are not approved for retrofit into existing household refrigerators designed for refrigerants other than hydrocarbons. New SNAP regulations have approved R-290, R-450A, and R-600a refrigerants for newly manufactured household refrigerators.

Recovery Techniques

When servicing a small appliance found to be leaking, it is not mandatory to repair the leak, but it is recommended to do so whenever possible. Before beginning a refrigerant recovery procedure, it is always necessary to know the type of refrigerant that is in the system to prevent the mixing of refrigerant in the recovery cylinder. It is unacceptable to mix any refrigerants in an appliance or recovery cylinder.

R-744, carbon dioxide CO_2, is a very high-pressure refrigerant and generally does not need to be recovered.

When R-410A or any other refrigerant is recovered from an appliance, it should be recovered into a recovery vessel that is clearly marked as to its content, to ensure that mixing of refrigerants does not occur, Figure 14.

Figure 14.
Recovery Tank.

Comparisons to a pressure-temperature chart are only valid to check for contaminated refrigerant if both the pressure and temperature of the refrigerant are stable and known. When checking for non-condensables inside a recovery cylinder containing one refrigerant, the technician must allow the temperature of the cylinder to stabilize to room temperature before taking a pressure reading for comparison using the pressure-temperature (PT) chart.

If a technician suspects that the refrigerant in a recovery cylinder to be contaminated, a pressure measurement should be taken and compared to a pressure-temperature chart. If the pressure-temperature relationship does not match, the refrigerant should be turned in to a reclamation facility. Some reclamation facilities may refuse to accept mixed refrigerants or charge extra for processing or destroying it.

The service aperture or process stub typically used on small appliances is a straight piece of tubing that is entered using a piercing access valve to add or remove refrigerant from a small appliance. When installing any type of access fitting onto a sealed system the fitting should be leak tested before proceeding with recovery. Because solderless type piercing valves tend to leak over time, they should not remain on the refrigeration systems after completion of the repair.

A passive, system-dependent, recovery process for small appliances can capture refrigerant in a non-pressurized container. When using passive recovery devices on small appliances with non-operating compressors it is essential to take measures to help release trapped, regulated refrigerant from the compressor oil during refrigerant recovery. It is helpful to heat (with a heat gun) and strike the compressor with a rubber mallet when recovering refrigerants into a non-pressurized container from a small appliance, such as a household refrigerator, with an inoperative compressor. Doing so will help free refrigerant that might be trapped under the oil in the compressor crankcase.

When using the system-dependent (passive) recovery process and the compressor does not run, both the high and low side of the system must be accessed for refrigerant recovery.

Both low and high-side access valves should be installed when recovering refrigerant from a household refrigerator or any other appliance to improve the speed of recovery. To ensure that all CFC, HCFC, or HFC refrigerant has been removed from a frost-free refrigerator, the defrost heater can be energized to increase the refrigerant's temperature and vaporize any liquid. An appliance with an operating compressor and completely restricted capillary tube or metering device requires only one access valve on the high side of the system. If the pressure in an appliance is 0 psig after installing and opening a piercing access valve, the recovery procedure cannot begin.

When using a self-contained (active) recovery device and there is a large volume of non-condensable gas in the refrigerant being recovered, excessive pressure can occur on the high side of the recovery machine and in the recovery cylinder. Non-condensable gas in the refrigerant can be checked for by using a pressure-temperature (PT) chart. The approximate pressure of an R-134a cylinder that does not contain any non-condensable impurities and is stored in a room at 80°F is equal to a pressure of 87 psig. As long as there is liquid refrigerant present and no impurities in the cylinder, the pressure and temperature should match what is shown on the PT chart.

During system recovery and/or repair, a pungent odor detected from the refrigerant normally indicates a compressor burn-out. The technician should watch for signs of oil contamination when recovering refrigerant from a system that experienced a compressor burn-out as the system will have to be flushed if contaminants are present. When nitrogen is used to pressurize or blow debris out of the system, the nitrogen may be vented to the ambient air.

Safety

Household appliances charged with hydrocarbon (HC) refrigerants require permanent safety markings on or near any evaporators, exposed refrigerant tubing, the machine compartment, and exterior of the refrigerator.

When recovering refrigerants, safe work practices include wearing safety eyewear and butyl-lined gloves when connecting/disconnecting hoses. When using nitrogen to service, leak test, or purge a refrigeration system, the nitrogen tank should be equipped with a regulator.

Because refrigerants are heavier than air and displace oxygen, a very large refrigerant leak can cause suffocation. If a large leak of refrigerant occurs, such as from a filled cylinder in an enclosed area, and no self-contained breathing apparatus is available, vacate and naturally ventilate the area. At high temperatures, CFC and HCFC refrigerants can decompose into phosgene gas, hydrochloric, and/or hydrofluoric acids.

Type II

Technicians maintaining, servicing, repairing or disposing of medium, high, and very high-pressure appliances, except small appliances as defined in Type I and motor vehicle air-conditioning systems (MVAC), must be certified as a Type II Technician or a Universal Technician.

System Classifications

Air-conditioning appliances used to supply cooling in order to control heat and/or humidity in occupied facilities is defined by EPA as Comfort Cooling for the leak repair category. Residential, office, and commercial buildings are included in this category. Appliances (equipment) used to preserve products, as in the retail food and cold storage warehouse sectors are classified by EPA as Commercial Refrigeration.

When an appliance has a dual use, the applicable maximum leak rate is determined by the highest percentage of use. For example; If a company uses 55% of its systems cooling capacity to one of its manufacturing lines or refrigeration and 45% of its cooling capacity to air condition the office spaces, it is considered Industrial Process Refrigeration (IPR) for the leak rate category.

Determining System Charge

Use the charge stated on the equipment nameplate to determine normal charge for leak rate calculations when the system is a packaged system charged at the factory. The charge in field-piped split-systems can be calculated using the factory charge plus the charge added for piping and accessories. The leak rate of an appliance can be calculated when topping off or recharging a system due to a loss of charge.

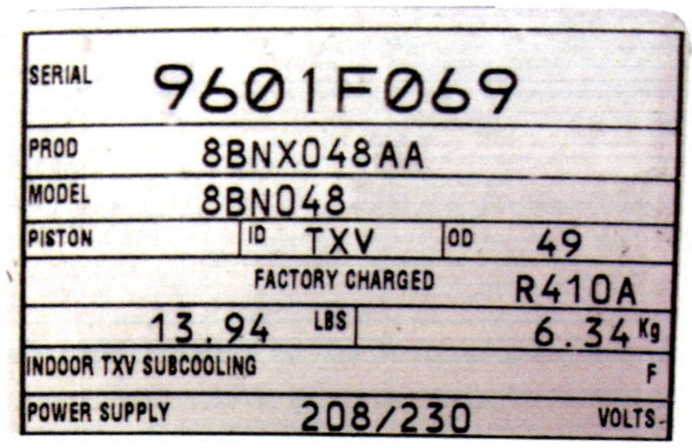

Figure 15. Data Plate.

The unit nameplate is the easiest way to check the type of refrigerant used in a packaged rooftop unit, Figure 15. In addition, the unit's nameplate will indicate the system's total charge. The best method to charge a system that has both an air-cooled condenser and evaporator with a nameplate charge of 80 pounds is a liquid charge by weight through the liquid-line service valve.

Type II Leak Repair Requirements

The EPA's leak repair requirements apply to all appliances using a regulated refrigerant. Once the threshold leak rate has been exceeded on any system with a charge size of 50 pounds or greater, the owner must repair the appliance enough to bring the leak rate below the threshold. If the owner cannot, or chooses not to, bring the leak rate below the threshold, the appliance must be retrofitted or retired.

Note: Leak repair requirements are the same for Type II & III.

As of January 1, 2019 the maximum leak rates are:

- 10% of the charge per year when used for comfort cooling.
- 20% of the charge per year when used for commercial refrigeration.
- 30% of the charge per year when used for industrial process refrigeration.

Leak Repair Time Frames

If an appliance with 50 or more pounds of a regulated refrigerant has exceeded the threshold leak rate, the equipment owner or operator has 30 days to repair the appliance so that it leaks below the threshold limit and conduct the initial verification test, unless EPA grants additional time. If an appliance with a charge of 200 pounds or more of a regulated refrigerant passes the initial verification leak test, a follow-up verification test must be conducted within 10 days. Not having a certified service technician available for service **cannot** be used as a reason to extend the appliance leak repair deadline.

If after repairing a leak in an HCFC rack refrigeration system used in a supermarket, the system fails the initial leak verification test, the technician has a 30-day window to make additional repairs or develop a retrofit plan.

Assuming no extensions apply, a cold storage warehouse operating an appliance charged with R-22 or any other regulated substance leaking above the threshold rate, can continue to operate the appliance without repair for 12 months before it must be retrofitted or retired. If that system leak rate exceeds 125% of its full charge in a calendar year, then a report must be submitted to EPA describing the efforts to identify and repair system leaks. Further action must be taken by repairing, retiring, mothballing, or retrofitting the appliance.

Note: An appliance using an ozone-depleting refrigerant does not need to be removed or mothballed by storing in a warehouse at the facility to extend the leak repair deadline. The repair deadline can be extended by temporarily shutting down the appliance, recovering the refrigerant from an isolated section or the entire system to at least atmospheric pressure.

The owner or operator has 18 months to retrofit or retire a leaking appliance if the replacement appliance will use a refrigerant exempt from the venting prohibition.

Section 608 Leak Repair Regulations

The following chart summarizes the repair regulations as outlined above.

Note: Leak repair requirements are the same for Type II & III.

Section 608 Leak Repair Regulations

The updated Section 608 regulations include new leak inspection and verification test requirements for owners/operators that affect technicians as of January 1, 2019:

Leak inspections are required for appliances that have exceeded the maximum allowable leak rate, according to the schedule below. All visible and accessible components of an appliance must be inspected, using a method or methods that are appropriate for that appliance.

Equipment	Full Charge	Frequency of Required Leak Inspections After Repair
Commercial Refrigeration and Industrial Process Refrigeration	More than 500 pounds	Once every three months until the owner/operator can demonstrate through leak rate calculations that the leak rate has not exceeded 20% (commercial refrigeration) or 30% (IPR) for four quarters in a row.
	50 to 500 pounds	Once per calendar year until the owner/operator can demonstrate through the leak rate calculations that the leak rate has not exceeded 20% (commercial refrigeration) or 30% (IPR) for one year.
Comfort Cooling	50 or more pounds	Once per calendar year until the owner/operator can demonstrate through the leak rate calculations that the leak rate has not exceeded 10% for one year.

Initial and follow-up verification tests of leak repairs are required for appliances that exceed the applicable leak rate. The verification tests must demonstrate that leaks were successfully repaired.

*__An initial verification test__ must be performed before any additional refrigerant is added to the appliance.

**__A follow-up verification__ test must be performed only after the appliance has returned to normal operating characteristics and conditions. There is no minimum timeframe.

***Technicians are required to** provide service invoices and records of leak inspections or verification tests to owners/operators.*

Example: Based on the Section 608 Leak Repair Regulations chart, an industrial process refrigeration (IPR) system containing 1000 lbs. or more of R-22, that has exceeded the applicable leak rate threshold, must be inspected for leaks every 3 months until verification that leak rate has not been exceeded for one year after repair.

Record Keeping

The records of leak inspections, initial leak verification, and follow-up verification tests must be kept on file for three (3) years. The owner and/or operator of the equipment is responsible for keeping the records.

Leak Detection

After assembly and installation of a split-system, the unit should first be pressurized with an inert gas, such as nitrogen (may also be referred to as dry-nitrogen), and leak checked. When absolutely necessary, dry-nitrogen with a trace amount of the system's design refrigerant should be used for leak detection. For example, an R-407C system should be leak checked with pressurized dry-nitrogen and a trace amount of R-407C.

When first inspecting an operating hermetic system known to be leaking, you should look for traces of oil. Oil traces on the rotating shaft seal of a belt-driven compressor is an indication of a leak, especially if the compressor has not been used for several months. To pinpoint refrigerant leaks, testing with soap bubbles is best.

A refrigerant leak in a high-pressure capillary tube system should be suspected if the system operates with excessive superheat.

Recovery Techniques: Before You Recover

Although those who service Type II, stationary refrigeration appliances must own active recycling and recovery equipment, it is no longer required that the purchase of the equipment be reported to the EPA.

Before you recover refrigerant, determine the maximum amount of refrigerant that is contained in the system. This will be the system's total charge. If the total charge is greater than 15 pounds, system-dependent, or passive, recovery techniques **cannot** be used as EPA regulations prohibit their use when the appliance contains over 15 pounds of refrigerant.

Before using a recovery unit to remove a charge, the service valve positions and recovery oil level should be checked. When recovering a different refrigerant type than the last refrigerant recovered using that specific recovery equipment, remove any refrigerant remaining in the recovery unit and change the unit's filter. The recovered refrigerant may have contained impurities of acids, moisture, and oil.

The gauge hoses used to connect the recovery unit to the tank (and for all field service work) are required to have self-sealing connectors or hand valves to minimize refrigerant release when the hoses are connected or disconnected.

Before transferring refrigerant to an empty cylinder, the cylinder should be evacuated to reduce the pressure and remove non-condensables.

Before recovering refrigerant in a parallel piped compressor system, make certain that the equalization connections between compressors are closed. Recovery machines that utilize a water-cooled condenser are generally connected to the municipal water supply.

Recovery Techniques: Enhancing the Process

To reduce recovery time, connect one of the service hoses to the liquid line and recover refrigerant in the liquid phase or remove the refrigerant from the condenser outlet when the condenser is below the receiver. After the liquid refrigerant has been recovered, the remaining vapor is removed by the recovery system. Recovering refrigerant in the vapor phase, although slower than liquid recovery, will minimize oil loss from the appliance. Refrigerant that has been recovered from an air-conditioning system and held in a refillable cylinder may be charged back into the system after the repair. If, however, the appliance is being disposed of, it is important to know that all HVAC service companies must keep records for three (3) years for each appliance disposed of if the system's refrigerant charge is between 5 and 50 pounds.

Cooling the recovery cylinder will lower its pressure and speed up the recovery process or removal of refrigerant charge from a system. In addition, heating the system will also speed up the recovery process as this will increase the pressure differential between the tank and the system. As the amount of refrigerant in the appliance drops, be careful to closely monitor the temperature of the recovery unit's compressor. This is because the hermetic compressor motors on recycling or recovery machines rely on the flow of refrigerant through the compressor for cooling and may overheat when drawing deep vacuums. In addition, the motor winding of a hermetic refrigeration compressor could be damaged if energized under a deep vacuum, so care should be taken to avoid energizing these compressors when refrigerant is being recovered

Recovery Techniques: Required Evacuation Levels During Recovery

The required levels of evacuation during recovery are summarized in the following chart:

Required Levels of Evacuation During Recovery for Type II Appliances [Except for small appliances, MVACs, and MVAC-like appliances]	
Using recovery and/or recycling equipment manufactured or imported after November 15, 1993	
Very high-pressure appliance	0" Hg
High-pressure appliance, or isolated component of such appliance, with a full charge of less than 200 lbs. of refrigerant.	0" Hg
High-pressure appliance, or isolated component of such appliance, with a full charge of 200 lbs. or more of refrigerant.	10" Hg
Medium-pressure appliance, or isolated component of such appliance, with a full charge of less than 200 lbs. of refrigerant.	10" Hg
Medium-pressure appliance, or isolated component of such appliance, with a full charge of 200 lbs. or more of refrigerant.	15" Hg

Example 1: Before making a major repair to an R-410A (a high-pressure refrigerant) system that contains more than 200 pounds of refrigerant, the technician must evacuate the appliance (or component thereof) to a vacuum level of at least 10 inches of Hg vacuum.

Example 2: After reaching the required recovery vacuum level on an appliance wait for a few minutes to see if the system pressure rises indicating moisture or a leak. For example, when changing out the compressor of a system containing 40 pounds of R-404A (a high-pressure refrigerant), evacuate the isolated section of the system to 0 psig and if system pressure does not rise after a few minutes, remove the compressor.

Example 3: If leaks in an appliance containing a regulated refrigerant make recovery to the prescribed level unattainable, the evacuation levels in the chart can be ignored, as the system refrigerant only needs to be recovered until the system's internal pressure reaches atmospheric pressure.

Note: *Depending on the question, the EPA exam may use the word* ***EVACUATION*** *in place of* ***Dehydration*** *or in other questions in place of* ***Recovery***.

Refrigeration System Liquid-Line Accessories

The refrigerant receiver will typically be found in the liquid line, at the outlet of the condenser, when the system uses a thermostatic expansion valve as its metering device. When the system is operating, the state of refrigerant leaving the receiver is a high-pressure liquid. Prior to opening a system with a liquid-line receiver for repair, the system's refrigerant should be pumped down and isolated in the receiver whenever possible to eliminate the need to recover the refrigerant from the system.

A moisture indicating sight glass is used for checking refrigerant for moisture. The sight glass is located in the liquid line, immediately downstream, after, the liquid-line filter drier. Anytime a system is opened for service, the liquid-line filter drier should be replaced. The filter drier is used to remove moisture from the refrigerant in a system.

Note: Isopropyl alcohol should be used to remove ice from sight glasses or viewing glasses.

Refrigeration System Suction Line and Compressor Accessories

In the suction line, at the outlet of the evaporator, suction line accumulators can be found. They are located between the evaporator and the compressor. Accumulators prevent any liquid refrigerant that may flow from the evaporator from flowing into the compressor. The normal position of the suction service/shutoff valve is back-seated. In the back-seated position, the gauge port is closed off so the manifold hose can be removed from the valve without releasing refrigerant. On the high side of the compressor, the discharge valve can be found. As with the suction service valve, the normal operating position is back-seated. Under no circumstances should the discharge service valve be closed on an operating reciprocating compressor. Doing so can result in compressor damage as well as severe personal injury. To help prevent compressor damage, refrigerant systems must be protected by a high-pressure relief device which may be in the compressor or external relief valve. When using multiple pressure relief valves they can only be installed in parallel.

Crankcase heaters are used on refrigeration and air-conditioning compressors to reduce the amount of refrigerant in the oil and to reduce oil foaming which usually occurs in the compressor crankcase. Should the compressor burn-out, an oil sample should be taken to test for the presence of acid in the system. Compressor replacement, as well as the replacement of an evaporator coil, condenser coil, or other system heat exchanger, is considered to be a "major" repair as defined by EPA regulations.

System Evacuation

A deep vacuum is measured in microns. A vacuum level of 500 microns should be achieved to ensure all moisture and non-condensables have been removed from a refrigeration system. When evacuating a system with large amounts of moisture it may be necessary to break the vacuum by increasing the pressure with nitrogen to counteract freezing of the moisture. Water in the system can freeze when using too large of a vacuum pump to evacuate a system. If non-condensables, such as air or dry-nitrogen, remain in a refrigeration system, the discharge pressure of the system will be high.

Notes: Nitrogen tanks are under high pressures, so the nitrogen cylinders require a pressure regulator to reduce the pressure to a safe level.

On a refrigeration system, never install multiple pressure relief valves in series.

System Charging and the Pressure-Temperature Chart

In an R-410A chilled-water system, vapor is charged from a vacuum level to a pressure of approximately 110 psig, which corresponds to a temperature of 36°F, to prevent water freezing. Similarly, the saturation pressure of an R-410A machine at (80°F) room temperature, while the machine is idle, is 238 psig when referring to a PT chart. An R-134a chiller should be initially charged with vapor to a pressure of approximately 30 psig, which corresponds to a temperature of 35°F. The evaporation temperature of R-134a at 0 psig is -15°F. For R-123, a low-pressure refrigerant that is used in low-pressure appliances, the saturation pressure at 35°F is 19.5"Hg vacuum (Refer to the Core section for information on how to use the PT chart). ASHRAE Standard 15 requires the use of sensors to detect the presence of refrigerant with few exceptions. ASHRAE Standard 15, for example, does not require an equipment room refrigerant detector for R-717 provided the mechanical ventilation system in the machinery room is run continuously.

Type III

Technicians maintaining, servicing, repairing, or disposing of low-pressure appliances must be certified as a Type III Technician or a Universal Technician. Most low-pressure appliances are used for comfort cooling and are subject to a lower allowable leak rate.

Leak Detection

Because a low-pressure system operates below atmospheric pressure (in a vacuum), leaks in the gaskets or fittings will cause air and moisture to enter the system.

Low-pressure chillers require purge units because they operate below atmospheric pressure, Figure 16. The primary purpose of a purge unit is to remove non-condensables that leak into a low-pressure chiller system. The purge unit gets its suction from a connection on top of the condenser where most non-condensables accumulate. After separating the air from the refrigerant, the purge unit sends the refrigerant back to the evaporator. High head pressure indicates air in a low-pressure system and that the purge unit may not be operating properly. Excessive running of a purge system on a low-pressure chiller generally indicates a leaking system. High-efficiency purge units will discharge a low percentage of refrigerant with the air they remove.

Figure 16. Purge Unit.

A hydrostatic tube test kit is used to determine if a tube is leaking. When leak testing a low-pressure centrifugal chiller with nitrogen, the maximum test pressure is 10 psig. Leak testing a low-pressure refrigeration system with nitrogen in excess of 10 psig could cause the rupture disc to fail. Charged low-pressure refrigeration machines may be most efficiently leak checked by adding heat with circulated hot water or heating blankets. To reduce refrigerant loss from the purge unit on an R-123 chiller, always leak test and repair the chiller. A low-pressure refrigeration system could have a leak in the condenser or chiller barrel tubes if excessive moisture continuously collects in the purge unit.

To prevent air accumulation into an idled low-pressure refrigeration system, maintain system pressure slightly above atmospheric pressure.

Moisture most frequently enters the refrigerant system in low-pressure chillers through leaks from areas with gaskets or fittings. To check for a refrigerant leak to the water side of a chiller, place a leak detector probe in the drain valve opening of the water box with the water removed. Drain the water sides of the evaporator and condenser prior to recovering refrigerant from a chiller suspected of having tube leaks.

Low-pressure refrigeration systems with open-drive type compressors are particularly susceptible to leaks from the shaft seal. According to ASHRAE Guideline 3, the system should be checked for leaks if, during a standard vacuum test, the pressure in a system rises from 1 mm Hg to a level above 2.5 mm Hg.

Leak Repair Requirements

The EPA's leak repair requirements apply to all appliances using a regulated refrigerant. Once the threshold leak rate has been exceeded on any system using a regulated refrigerant with a charge of 50 pounds or greater, the owner must repair the appliance to bring the leak rate below the threshold, retrofit, or retire the appliance.

As of January 1, 2019, the applicable maximum leak rates are:
- 10% of the charge per year when used for comfort cooling.
- 20% of the charge per year when used for commercial refrigeration.
- 30% of the charge per year when used for industrial process refrigeration.

Chillers generally belong to the EPA leak repair category for comfort cooling. When used in multiple categories, the applicable maximum leak rate is determined by the highest percentage of use, Figure 17.

Figure 17. Low-Pressure Chiller.

Example: If a company diverts 55% of its chiller's cooling capacity to one of its manufacturing lines and 45% of its cooling capacity to air condition the office spaces, the chiller is considered in the industrial process refrigeration (IPR) leak rate category.

Use the charge stated on the equipment nameplate to determine a chiller's normal charge for leak rate calculations. The leak rate of an appliance using a regulated refrigerant must be calculated when topping off or recharging a system due to a loss of charge.

If an appliance with 50 or more pounds of a regulated refrigerant charge has exceeded the threshold leak rate, the owner or operator has 18 months to retrofit or retire the appliance if the replacement appliance will use a refrigerant exempt from the venting prohibition.

Not having a certified service technician available for service cannot be used as a reason to extend the appliance leak repair deadline.

System mothballing does not require removing the appliance and storing it in a warehouse at the facility.

If an appliance containing 50 or more pounds of a regulated refrigerant has exceeded the threshold leak rate, the owner or operator has 30 days to repair the appliance so that it leaks below the threshold and conduct the initial verification test.

If a chiller with a charge of 200 pounds of an ozone-depleting refrigerant has passed the initial verification leak test, a follow-up verification test must be conducted within 10 days.

The owner and/or operator of the equipment is responsible for and must keep records of all leak inspections, initial verification, and any completed leak repair follow-up verification tests for three (3) years.

Recovery Techniques

The high-pressure cut-out control on a recovery unit used for evacuating/recovering the refrigerant from a low-pressure chiller is typically set for 10 psig. Due to the volume of liquid refrigerant in large systems, refrigerant removal from a low-pressure system starts with liquid removal followed by recovery of the vapor refrigerant. After all the liquid has been removed, an average 350-ton, R-123 chiller, at 0 psig, would still have approximately 100 pounds of refrigerant in the vapor state.

The use of a severely oversized vacuum pump could cause trapped water to freeze when evacuating/dehydrating a system. During evacuation/dehydration of a system containing large amounts of moisture, it may be necessary to break the vacuum using nitrogen. Increasing the pressure with nitrogen prevents the moisture from freezing and increases the speed of dehydration.

Figure 18. Rupture Disc.

A water-cooled recovery unit will allow for faster recovery of large quantities of refrigerant. Water is circulated through a chiller during refrigerant evacuation/recovery to prevent freezing of water in the appliance.

The system's water pumps, recovery compressor, and recovery condenser water should all be on during vapor or liquid removal from a low-pressure refrigeration system. Raising the temperature of the system or the room in which a low-pressure chiller is located will also result in faster refrigerant recovery.

When removing refrigerant oil, it should be heated to 130°F as less refrigerant will be contained in the oil at the higher temperature. Be careful when heating the system, since the rupture disc on a recovery vessel for low-pressure refrigerants relieves at 15 psig, Figure 18.

Recovery Requirements

After reaching the required recovery vacuum on a low-pressure appliance wait a few minutes to see if the system pressure rises before charging. When leaks in the appliance make evacuation/recovery to the prescribed level unattainable, do not proceed with the recovery. Recovery and recycling equipment must be labeled as certified to meet EPA's requirements.

Replacing or repairing a compressor, condenser, or evaporator is considered a major repair. When making a non-major repair, warming the liquid refrigerant in the system with hot water or warming blankets will bring the system pressure out of a vacuum.

Before disposing of a low-pressure appliance, the refrigerant must be recovered or evacuated to 25 mm of Hg absolute. Recovery records for disposed appliances with 5-50 lbs. of refrigerant must be kept for three (3) years.

Recharging Techniques

Before evacuation/recovery or dehydration, an oil sample should be taken if the unit has had a compressor burn-out. Also, when evacuating/dehydrating a system that contains a large amount of moisture, triple evacuate/dehydrate the system using dry-nitrogen to break the vacuum. Increasing the pressure with nitrogen will prevent the moisture from freezing.

After servicing and evacuating/dehydrating a chiller, refrigerant vapor is re-introduced into the chiller to a pressure which corresponds to its saturation temperature above freezing. This will prevent liquid refrigerant charged into the system from freezing water in the heat exchanger tube bundle. For example; charging liquid R-245fa into a low-pressure refrigeration system in a vacuum greater than 18 inches Hg vacuum will cause the system water to freeze. At 18 inches Hg vacuum, R-245fa has a saturation temperature well below -40°F. When reviewing the saturated conditions on a PT chart, an evacuated/dehydrated water-cooled chiller system with R-123 should have a vapor pressure above 20 inches Hg vacuum before adding liquid refrigerant. The liquid refrigerant is charged into the system using the evaporator charging valve, which is the lowest access point on a low-pressure appliance.

Note: The relationships found on pressure-temperature charts will vary slightly from one manufacturer or publisher to the next. The pressures are normally indicated in psig. To convert from absolute pressure (psia) to gauge pressure (psig) subtract 14.7 from the absolute pressure. Therefore, 14.7 psia equals 0 psig.

Safety

When the TLV-TWA (Threshold Limit Value - Time Weighted Average) is exceeded, ASHRAE Standard 15-2013 requires that each machinery room activate an alarm and a mechanical ventilation system.

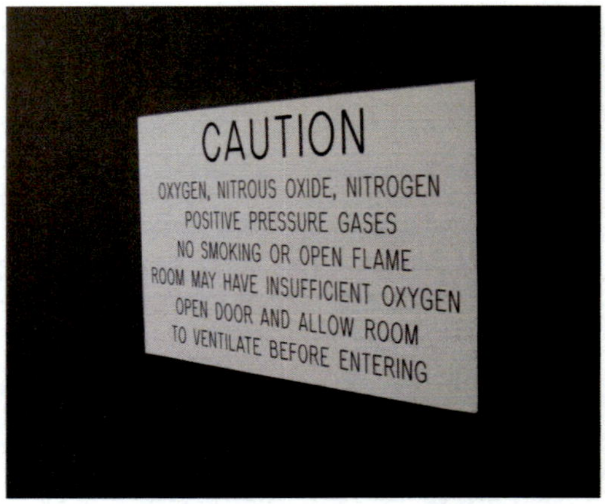

Figure 19. Equipment Room Wall Plate.

ASHRAE Standard 15-2013 also requires the use of room sensors and alarms to detect refrigerant leaks in all refrigerant safety groups, Figure 19. R-123 refrigerant is classified as B1 (non-flammable, high toxicity) and refrigerant R-1233zd is classified A1 (non-flammable, low-toxicity) under Standard 34 code group. The use of room sensors and alarms to detect refrigerant leaks in all refrigerant safety groups is required not only because of refrigerant characteristics, but because refrigerants are heavier than air and can displace oxygen.

Gloves and safety goggles should be worn when working with liquid refrigerant for low-pressure systems.

The discharge from a rupture disc or pressure relief valves should be piped outdoors to prevent venting into the machine room. All pressure relief valves should be piped or installed in parallel, **never** in series.

A 15 psig rated rupture disc is located on the evaporator of a centrifugal chiller. When charging, refrigerant is introduced through the evaporator charging valve, make sure not to over-pressurize the system.

Note: Isopropyl alcohol should be used to remove ice from sight glasses or viewing glasses.